The Hidden Universe: A Journey through Dark Matter and Dark Energy for Everyone

Kayson Rory

Copyright © [2023]

Title: The Hidden Universe: A Journey through Dark Matter and Dark Energy for Everyone
Author's: Kayson Rory

This book was printed and published by [Publisher's: **Kayson Rory**] in [2023]

ISBN:

TABLE OF CONTENT

Chapter 1: Introduction to the Hidden Universe 08

What is Dark Matter?

The Discovery of Dark Matter

The Enigma of Dark Energy

The Importance of Dark Matter and Dark Energy

Chapter 2: The History of Dark Matter 16

Early Theories and Concepts

The Galactic Rotation Problem

The Bullet Cluster Observation

The WIMP Hypothesis

Chapter 3: The Hunting Grounds of Dark Matter   24

Underground Experiments

Direct Detection Methods

Indirect Detection Methods

The Search for Axions

Chapter 4: Unraveling the Nature of Dark Matter   32

The Standard Model and Beyond

Supersymmetry and Weakly Interacting Massive Particles (WIMPs)

Other Theoretical Candidates for Dark Matter

The Role of Particle Colliders in Dark Matter Research

Chapter 5: The Mysteries of Dark Energy 40

The Accelerating Expansion of the Universe

The Cosmological Constant

Alternative Theories of Dark Energy

Probing Dark Energy with Supernovae and Cosmic Microwave Background

Chapter 6: The Cosmic Web: Dark Matter's Influence on Structure Formation 48

The Birth of the Universe

Gravitational Clustering

Simulations and Observations of Large-Scale Structure

Dark Matter Halos and Galaxy Formation

Chapter 7: The Future of Dark Matter and Dark Energy Research   56

Advancements in Experimental Techniques

The Role of Space Missions

Multi-Messenger Astronomy and Dark Matter

The Quest for a Unified Theory of Everything

Chapter 8: Implications and Applications of Dark Matter and Dark Energy   64

Dark Matter in the Milky Way and Beyond

Dark Energy's Influence on the Fate of the Universe

Implications for Astrophysics and Cosmology

The Potential for Technological Breakthroughs

Chapter 9: The Human Connection: Why Dark Matter and Dark Energy Matter to Everyone 73

The Quest for Knowledge and Understanding

Dark Matter and Dark Energy in Popular Culture

Inspiring the Next Generation of Scientists

The Role of Citizen Science in Dark Matter and Dark Energy Research

Chapter 10: Conclusion 81

Recap of Key Discoveries and Concepts

Looking Ahead: The Continued Exploration of the Hidden Universe

Final Thoughts and Reflections.

Chapter 1: Introduction to the Hidden Universe

What is Dark Matter?

In the vast expanse of the universe, there exists a mysterious and elusive substance known as dark matter. This enigmatic entity has captivated scientists and astronomers alike for decades, as its presence holds profound implications for our understanding of the cosmos. But what exactly is dark matter, and why is it so crucial in the realm of particle physics?

Dark matter refers to a type of matter that does not interact with light or other forms of electromagnetic radiation, making it invisible to our conventional detection methods. This invisible substance constitutes about 85% of the matter in the universe, dwarfing the ordinary matter that we can observe and interact with. Its existence was first hypothesized by the Swiss astronomer Fritz Zwicky in the 1930s, who noticed discrepancies between the visible matter and the gravitational effects observed in galaxy clusters.

So, if dark matter is invisible, how do we know it exists? Scientists have employed various ingenious methods to indirectly detect the presence of dark matter. One such method is studying the movements of stars and galaxies. By observing their motions, scientists can infer the gravitational effects exerted by the invisible dark matter, which acts as an invisible cosmic glue, holding these massive structures together.

Particle physics plays a crucial role in understanding dark matter. Scientists believe that dark matter is composed of exotic particles that do not belong to the known roster of elementary particles. These particles do not interact with light, making their detection immensely challenging. However, through sophisticated experiments conducted in underground laboratories and particle colliders, scientists are relentlessly searching for these elusive particles.

The discovery of dark matter would revolutionize our understanding of the universe, shedding light on the nature of gravity, the formation of galaxies, and the evolution of our cosmos. It may also provide answers to questions about the elusive dark energy that is driving the accelerated expansion of the universe. Understanding dark matter is crucial in unraveling the mysteries of our universe and may even lead to groundbreaking technological advancements in the future.

In conclusion, dark matter remains a silent protagonist in the cosmic drama, quietly shaping the universe as we know it. Its elusive nature challenges our understanding of the fundamental forces and particles that govern the universe. As scientists continue to delve deeper into the mysteries of the hidden universe, the search for dark matter continues to captivate our imaginations, promising a journey of discovery that will unlock the secrets of our existence.

The Discovery of Dark Matter

In the vast expanse of the universe, there exists a mystery that has baffled scientists for decades - the enigma of dark matter. This invisible substance, which does not interact with light, has eluded our detection and comprehension, leaving us with more questions than answers. In this subchapter, we will embark on a journey through the captivating story of the discovery of dark matter, a truly groundbreaking achievement in the field of particle physics.

The story begins in the early 20th century when astronomers noticed discrepancies between the observed gravitational forces and the visible matter in galaxies. Swiss astronomer Fritz Zwicky was the first to propose the existence of unseen matter that affected the motion of galaxies. However, it wasn't until the 1970s that the search for dark matter gained momentum.

Two independent teams, led by American astronomers Vera Rubin and Kent Ford, set out to study the rotation curves of galaxies. Their observations revealed that the outer regions of galaxies were rotating at unexpectedly high speeds, defying the laws of gravity. This strange phenomenon could only be explained by the presence of invisible matter - dark matter.

As excitement grew within the scientific community, physicists turned their attention to particle accelerators, such as the Large Hadron Collider (LHC), in search of dark matter particles. The LHC, nestled deep beneath the Swiss-French border, became the stage for groundbreaking experiments. Although no direct detection of dark matter particles has been achieved so far, these experiments have provided valuable insights into the nature of this elusive substance.

One of the leading candidates for dark matter particles is the Weakly Interacting Massive Particles (WIMPs). These hypothetical particles, which interact extremely weakly with ordinary matter, are believed to have played a crucial role in the formation and evolution of galaxies. Scientists continue to refine their theories and develop increasingly sensitive detectors in the quest to unveil the true identity of dark matter.

The discovery of dark matter has not only revolutionized our understanding of the universe but has also opened up new avenues for scientific exploration. Its existence prompts us to question the very foundations of our understanding of physics and challenges us to unlock the secrets of the hidden universe.

In conclusion, the discovery of dark matter stands as one of the most significant achievements in the field of particle physics. From the early observations of gravitational anomalies to the ongoing experiments at particle accelerators, the pursuit of this elusive substance has captivated scientists and transformed our understanding of the universe. As we delve deeper into the mysteries of dark matter, we inch closer to unraveling the secrets of our hidden universe.

The Enigma of Dark Energy

In the vast expanse of the cosmos, there exists a mysterious force that defies our understanding - dark energy. This enigmatic entity has captivated the minds of scientists and perplexed even the most brilliant minds in the field of particle physics. In this subchapter, we embark on a journey through the depths of the hidden universe to unravel the secrets of dark energy.

Dark energy is a term used to describe the unknown force that is driving the accelerated expansion of our universe. Unlike the more familiar dark matter, which exerts gravitational forces on visible matter, dark energy operates on a much larger scale, influencing the very fabric of space-time itself. Its existence was first hypothesized when astronomers observed that galaxies were moving away from each other at an ever-increasing rate.

One of the greatest challenges in understanding dark energy lies in its elusive nature. Unlike particles that can be studied and observed, dark energy remains undetectable through conventional means. Its presence can only be inferred from its effects on the universe. Scientists have proposed various theories to explain its origin, ranging from the vacuum energy of empty space to the existence of a new subatomic particle yet to be discovered.

The implications of dark energy extend far beyond the realm of particle physics. It has the power to shape the fate of our universe. If dark energy continues to drive the accelerated expansion, it could ultimately lead to the "Big Rip," a cataclysmic event where galaxies, stars, and even atoms are torn apart.

While the enigma of dark energy continues to baffle scientists, its study has opened up new avenues of research and sparked groundbreaking discoveries. The quest to understand this mysterious force has led to advancements in our understanding of fundamental physics, challenging long-held beliefs and pushing the boundaries of human knowledge.

In this subchapter, we delve into the intricacies of dark energy, exploring the current theories and research efforts dedicated to unlocking its secrets. From the cosmic microwave background radiation to the observations of distant supernovae, we examine the tools and techniques used to probe the mysteries of the hidden universe.

"The Enigma of Dark Energy" is not just a subchapter for particle physicists, but for everyone curious about the nature of our universe. It invites readers on a journey of exploration, providing a glimpse into the cutting-edge research that seeks to unravel the mysteries of dark energy and shed light on the fundamental workings of our cosmos.

The Importance of Dark Matter and Dark Energy

In the vast expanse of the universe, there are hidden forces at play that have puzzled scientists and astronomers for decades. These forces are known as dark matter and dark energy, and they hold the key to understanding the true nature of our universe. In this subchapter, we will delve into the importance of dark matter and dark energy and explore their significance in the realm of particle physics.

Dark matter, as the name suggests, is a mysterious substance that cannot be directly observed or detected. However, its presence is inferred through its gravitational effects on visible matter. Scientists believe that dark matter makes up around 27% of the universe, while ordinary matter accounts for only 5%. This means that the majority of the universe is composed of an invisible substance that we are still struggling to comprehend.

But why is dark matter important? Well, its existence helps explain the structure and formation of galaxies. Without the gravitational pull of dark matter, galaxies would not have enough mass to hold their shape and would simply fly apart. Dark matter also plays a crucial role in the evolution of the universe, as it acts as the scaffolding upon which galaxies and galaxy clusters form.

Moving on to dark energy, it is an even more enigmatic force that permeates all of space. Unlike dark matter, which acts as an attractive force, dark energy behaves as a repulsive force, causing the expansion of the universe to accelerate. This discovery, made in the late 1990s, was a groundbreaking revelation that earned the Nobel Prize in Physics in 2011.

Understanding dark energy is vital because it holds the key to the fate of our universe. If dark energy continues to drive the expansion, galaxies will eventually be torn apart, and the universe will become an ever-expanding void. On the other hand, if dark energy weakens or changes, gravity may take over, causing the universe to collapse in a Big Crunch. The nature of dark energy will ultimately determine the destiny of our cosmos.

In the realm of particle physics, dark matter and dark energy present profound challenges and opportunities. They push the boundaries of our understanding and compel scientists to develop new theories and experiments. By studying these invisible forces, we not only gain insight into the hidden workings of the universe but also deepen our knowledge of fundamental particles and their interactions.

In conclusion, the importance of dark matter and dark energy cannot be overstated. They hold the key to unraveling the mysteries of the universe and provide invaluable clues about the nature of particle physics. As we journey through the hidden universe, let us embrace the enigma of dark matter and dark energy and strive to uncover the secrets they hold.

Chapter 2: The History of Dark Matter

Early Theories and Concepts

In the vast expanse of the universe, there exists a mysterious realm that has captivated the minds of scientists and enthusiasts alike - the realm of dark matter and dark energy. These enigmatic entities have played an integral role in shaping the universe as we know it, yet their true nature remains elusive. To embark on a journey through this hidden universe, we must first understand the early theories and concepts that paved the way for our current understanding.

The exploration of dark matter and dark energy began with the birth of particle physics, a field that delves into the fundamental building blocks of the universe. Early on, scientists proposed various theories to explain the nature of these invisible forces. One of the earliest concepts was the existence of "invisible matter" that exerts gravitational influence on visible matter. This idea gained traction as astronomers observed galaxies rotating faster than expected, suggesting the presence of additional mass beyond what we can perceive.

As research progressed, scientists developed the theory of supersymmetry, which posits that every known particle has a yet-to-be-discovered counterpart. These hypothetical particles, known as supersymmetric particles or "sparticles," could potentially account for the missing mass in the universe. While the search for sparticles continues, it has yet to yield definitive evidence.

Another crucial concept in understanding dark matter and dark energy is the theory of cosmic inflation. Proposed in the early 1980s, this theory suggests that the universe underwent a rapid expansion in its infancy, leaving behind remnants of this inflationary period. These remnants, known as cosmic microwave background radiation, provide valuable insights into the early universe and its composition.

Furthermore, the discovery of the Higgs boson at the Large Hadron Collider (LHC) shed light on the fundamental mechanisms that govern particle interactions. The Higgs boson is associated with the Higgs field, which permeates all of space and gives particles mass. Understanding the Higgs boson and its interactions may provide clues about the nature of dark matter and dark energy.

As we delve deeper into the hidden universe, it is essential to grasp the early theories and concepts that have shaped our understanding thus far. From the existence of invisible matter to the search for supersymmetric particles and the theory of cosmic inflation, these ideas form the foundation of our quest for knowledge. With each new discovery, we draw closer to unraveling the mysteries of dark matter and dark energy, enlightening not only the field of particle physics but also the curious minds of every individual embarking on this cosmic journey.

The Galactic Rotation Problem

One of the most intriguing mysteries in the realm of particle physics is the Galactic Rotation Problem. This enigmatic phenomenon has puzzled scientists for decades, as it challenges our understanding of the laws of gravity and the composition of the universe.

When astronomers first began studying the rotation of galaxies, they expected to find a predictable pattern. According to Newton's laws of gravity, the outer regions of a galaxy should rotate at a slower velocity than the inner regions. However, careful observations revealed something entirely unexpected – the stars at the outskirts of galaxies were moving just as fast as those near the center. This anomalous behavior defied conventional explanations, leading researchers to dub it the Galactic Rotation Problem.

To unravel this cosmic conundrum, scientists turned to the concept of dark matter. Dark matter is a hypothetical form of matter that does not emit, absorb, or reflect light, making it invisible and difficult to detect. It is believed to make up around 85% of the total matter in the universe. The existence of dark matter was postulated to explain the gravitational effects observed in galaxies.

According to the dark matter theory, galaxies are embedded within massive halos of this elusive substance. These dark matter halos provide the extra gravitational pull needed to account for the rapid rotation of stars in the outer regions of galaxies. While dark matter itself remains elusive, its presence is inferred by its gravitational effects.

The Galactic Rotation Problem not only challenges our understanding of gravity but also sheds light on the existence of the equally perplexing dark energy. Dark energy is believed to be responsible for the accelerated expansion of the universe. Although it cannot be directly observed, its influence can be inferred from the observed behavior of galaxies and other celestial objects.

By unraveling the secrets of dark matter and dark energy, scientists hope to gain a deeper understanding of the fundamental laws of the universe. The Galactic Rotation Problem serves as a testament to the mysteries that lie within the hidden realms of particle physics. Exploring these mysteries not only expands our knowledge but also brings us closer to unraveling the secrets of the universe itself.

In conclusion, the Galactic Rotation Problem is an intriguing puzzle that has captivated the minds of scientists and astronomers. By delving into the depths of particle physics, we can begin to unravel the enigma of dark matter and dark energy, and ultimately gain a deeper understanding of the hidden universe that surrounds us all.

The Bullet Cluster Observation

In the vast expanse of the cosmos, scientists have always been intrigued by the mysteries that lie in the darkest corners of the universe. One such mystery that has captivated the minds of physicists and astronomers alike is the enigma of dark matter and dark energy. These invisible forces, which make up a staggering 95% of the universe, continue to baffle scientists, pushing them to explore deeper into the unknown.

One of the most compelling pieces of evidence for the existence of dark matter came from the observation of the Bullet Cluster. This celestial marvel, located over three billion light-years away, provided a unique opportunity to unravel the secrets of the universe.

The Bullet Cluster is actually a cluster of galaxies that collided with another cluster at an incredibly high speed. This collision caused a separation of the ordinary matter, such as gas and stars, from the dark matter. As a result, two distinct regions formed - one where the ordinary matter resides and another where the majority of the dark matter is located.

Through meticulous observations and analysis, scientists were able to measure the gravitational lensing effect caused by the dark matter in the Bullet Cluster. This effect occurs when the mass of an object bends the path of light passing near it. By studying the distortion of light from distant galaxies, researchers were able to indirectly detect the presence of dark matter.

The observations from the Bullet Cluster provided compelling evidence that dark matter exists and plays a crucial role in shaping the universe. It also revealed that dark matter interacts very weakly with ordinary matter, making it extremely difficult to detect using conventional methods.

This groundbreaking discovery opened up new avenues of research in the field of particle physics. Scientists began developing new technologies and experiments to directly detect dark matter particles. These experiments, conducted deep underground or in space, aim to identify the elusive particles that make up dark matter, offering a deeper understanding of the universe's hidden workings.

The Bullet Cluster observation not only shed light on the existence of dark matter but also deepened our understanding of the fundamental forces and particles that govern the universe. As we continue to explore the mysteries of the cosmos, this remarkable observation serves as a reminder that there is still much to uncover about the hidden universe of dark matter and dark energy.

The WIMP Hypothesis

In the vast expanse of the universe, there is something invisible that pervades every corner of space. It is a mysterious substance known as dark matter, and its presence has perplexed scientists for decades. To unravel this cosmic enigma, the world of particle physics has proposed a fascinating hypothesis called the WIMP hypothesis.

WIMP, which stands for Weakly Interacting Massive Particle, postulates the existence of a new type of particle that interacts weakly with ordinary matter. These particles, if they exist, would have a mass several times larger than that of a proton, yet they would barely interact with light or other electromagnetic radiation. This makes them incredibly difficult to detect, thus earning them the moniker "dark matter."

The WIMP hypothesis emerged as a potential solution to explain the gravitational effects observed in galaxies and galaxy clusters that cannot be accounted for by the visible matter alone. According to this hypothesis, dark matter particles would be abundant in the universe, forming invisible halos around galaxies and providing the gravitational glue that holds them together.

The search for WIMPs has been an ongoing endeavor, with scientists employing a range of ingenious detectors deep underground to shield against cosmic rays and other background noise. These experiments aim to catch a glimpse of the rare occasions when a WIMP interacts with ordinary matter, producing a signal that can be detected.

While no direct detection of WIMPs has been made so far, the WIMP hypothesis remains one of the most compelling explanations for the existence of dark matter. It offers a plausible solution to many astronomical observations that cannot be explained by the known particles of the Standard Model of particle physics.

Furthermore, the existence of WIMPs would have profound implications for our understanding of the universe. It could shed light on the nature of dark matter and provide insights into the formation and evolution of galaxies and the large-scale structure of the cosmos.

As we delve deeper into the hidden universe, the search for WIMPs continues. Scientists from around the world are collaborating to push the boundaries of knowledge and unravel the mysteries of dark matter. The WIMP hypothesis represents a crucial piece of the puzzle, offering a tantalizing glimpse into the true nature of the hidden universe that surrounds us all.

Chapter 3: The Hunting Grounds of Dark Matter

Underground Experiments

In the quest to unravel the mysteries of the universe, scientists and researchers have turned their attention to the depths of the Earth. Underground experiments have become a crucial component of particle physics, providing valuable insights into the nature of dark matter and dark energy. These experiments take place deep beneath the surface, shielded from cosmic rays and other interference, allowing scientists to study elusive particles and phenomena that would otherwise go undetected.

The underground environment offers unique advantages for particle physics experiments. By being shielded from cosmic rays, which constantly bombard the Earth from outer space, scientists can accurately measure the properties and interactions of particles without the interference caused by extraneous background radiation. This shields delicate experiments from unwanted noise, enabling researchers to study rare events with greater precision.

One of the most renowned underground experiments is the Large Underground Xenon (LUX) experiment, which was designed to detect dark matter particles. Located in the Sanford Underground Research Facility in South Dakota, LUX consists of a tank filled with liquid xenon, which is extremely sensitive to the interactions of dark matter particles. The experiment aims to directly detect and measure dark matter, which is believed to make up a significant portion of the universe's mass.

Another groundbreaking underground experiment is the Sudbury Neutrino Observatory (SNO) in Ontario, Canada. This experiment focuses on studying neutrinos, elusive particles that have negligible mass and interact weakly with matter. By utilizing a giant spherical tank filled with heavy water, SNO captures and analyzes neutrinos emitted by the Sun. This experiment provided crucial evidence for neutrino oscillation, a phenomenon that demonstrated neutrinos possess mass, challenging previous assumptions about their properties.

In addition to dark matter and neutrino experiments, underground facilities also play a vital role in the search for rare radioactive decays and other fundamental particle interactions. These experiments require an environment with extremely low levels of background radiation, which can be achieved deep underground. Facilities such as the Gran Sasso National Laboratory in Italy and the Homestake Mine in South Dakota house a range of experiments aiming to understand the fundamental building blocks of our universe.

Underground experiments have revolutionized our understanding of particle physics, shedding light on the enigmatic dark matter and the elusive neutrinos. Through their innovative designs and sophisticated technologies, these experiments offer unprecedented opportunities to explore the hidden universe. As we delve deeper into the mysteries of dark matter and dark energy, underground facilities will continue to play a crucial role in unraveling the secrets of the universe and expanding our knowledge of particle physics for everyone.

Direct Detection Methods

In the fascinating realm of particle physics, scientists are constantly exploring innovative ways to unravel the mysteries of the universe. One crucial aspect of this quest involves the detection of elusive entities known as dark matter and dark energy. Both of these enigmatic phenomena have mesmerized researchers for decades, and their discovery promises to revolutionize our understanding of the cosmos. In this subchapter, we delve into the intriguing realm of direct detection methods that scientists employ to search for dark matter particles.

Direct detection methods involve the direct observation of dark matter particles as they interact with ordinary matter. Over the years, scientists have developed ingenious techniques to capture these extremely rare interactions, hoping to shed light on the nature and properties of dark matter. The primary goal of direct detection experiments is to identify the weak signals left behind by dark matter particles passing through Earth.

One of the most popular direct detection methods involves the use of highly sensitive detectors buried deep underground. These detectors are shielded from cosmic rays and other background radiation that could interfere with the detection process. They are typically made of special materials, such as germanium or liquid xenon, which are capable of registering the faint signals produced by dark matter interactions.

When a dark matter particle collides with an atomic nucleus in the detector, it imparts a small amount of energy, resulting in a minute but detectable recoil. Scientists meticulously analyze these recoils, searching for characteristic patterns that differentiate them from background noise. By carefully studying the properties of these interactions, researchers can infer crucial information about the mass, velocity, and distribution of dark matter particles.

Another direct detection method involves the search for dark matter annihilation or decay products. Scientists look for the energetic particles or radiation that might be produced when dark matter particles collide and annihilate with each other. By studying the distribution and characteristics of these annihilation products, researchers can gain valuable insights into the abundance and behavior of dark matter.

Direct detection methods represent a crucial component of the quest to unravel the mysteries of dark matter and dark energy. Although the challenges are numerous, the relentless efforts of scientists worldwide continue to push the boundaries of our knowledge. With each new experiment and technological advancement, we come closer to unraveling the secrets of the hidden universe and understanding the fundamental nature of our existence.

Indirect Detection Methods

In the quest to unravel the mysteries of the universe, scientists have turned to various detection methods to study the elusive dark matter and dark energy. Indirect detection methods play a crucial role in our understanding of these enigmatic cosmic entities. By observing the effects and interactions of particles and radiation, scientists can infer the presence and properties of dark matter and dark energy, even though they cannot be directly observed.

One of the primary indirect detection methods used in particle physics is the measurement of cosmic rays. Cosmic rays are high-energy particles that originate from outer space and bombard our planet. By analyzing the composition and energy distribution of cosmic rays, scientists can gain insights into the presence and behavior of dark matter. Certain characteristics, such as an excess of positrons or gamma rays in cosmic rays, may indicate the annihilation or decay of dark matter particles.

Another powerful indirect detection method is the study of gravitational lensing. According to Einstein's theory of general relativity, massive objects can bend and distort the path of light. By observing these gravitational lensing effects, scientists can map the distribution of dark matter in the universe. Dark matter's gravitational influence on visible matter can be used as evidence for its existence and provide valuable information about its nature.

Astrophysical observations also contribute to indirect detection methods. By studying the motion of stars and galaxies, scientists can measure the gravitational forces acting on them. Deviations from the expected gravitational effects can be attributed to the presence of dark matter. These observations help refine our understanding of the distribution and density of dark matter in different regions of the universe.

Furthermore, indirect detection methods involve the search for exotic particles produced by the interactions of dark matter. Sophisticated detectors, such as the Large Hadron Collider, are used to recreate the conditions of the early universe and study the potential production of dark matter particles. The detection of these particles would provide direct evidence for dark matter and offer insights into its properties and interactions with ordinary matter.

Indirect detection methods have revolutionized our understanding of the hidden universe of dark matter and dark energy. They allow scientists to probe the invisible realms of particle physics and unravel the mysteries of the cosmos. By utilizing cosmic rays, gravitational lensing, astrophysical observations, and particle detectors, researchers can piece together the puzzle of dark matter and dark energy, bringing us closer to a comprehensive understanding of the universe we inhabit.

The Search for Axions

In the fascinating world of particle physics, scientists are constantly on the hunt for the smallest building blocks of our universe. These elusive particles hold the key to understanding the mysteries of the cosmos. One such particle that has captured the attention of researchers is the axion.

Axions are hypothetical particles that were first proposed in the late 1970s as a solution to a problem in quantum chromodynamics (QCD), a theory that describes the behavior of subatomic particles. They were initially postulated to explain why the strong nuclear force, which holds atomic nuclei together, does not violate certain symmetry principles.

However, it was soon realized that axions could also have profound implications for our understanding of dark matter. Dark matter, which makes up about 27% of the universe, is an invisible substance that does not interact with light or other forms of electromagnetic radiation. Its existence is inferred from its gravitational effects on visible matter.

Scientists believe that axions could be a major component of dark matter. Unlike other proposed candidates, such as weakly interacting massive particles (WIMPs), axions are incredibly light and interact even less with normal matter. This makes them extremely difficult to detect.

To search for axions, physicists have devised ingenious experiments that push the boundaries of technology. One such experiment is the Axion Dark Matter eXperiment (ADMX), which uses a high-quality magnetic field to convert axions into detectable microwave photons.

Another approach is the use of powerful lasers to convert axions into detectable particles. These experiments, such as the International Axion Observatory (IAXO), aim to create intense magnetic fields that can convert axions into photons that can be detected by sophisticated instruments.

The search for axions is not only about unraveling the mysteries of dark matter but also about pushing the boundaries of our knowledge of particle physics. If axions are discovered, it would be a groundbreaking moment in the field, opening up new avenues of research and potentially revolutionizing our understanding of the universe.

While the search for axions continues, scientists remain hopeful that these elusive particles will one day be found. Their discovery would not only provide insights into the nature of dark matter but also shed light on the fundamental workings of the universe.

In conclusion, the search for axions represents a fascinating quest within the realm of particle physics. These hypothetical particles, if discovered, could hold the key to unlocking the mysteries of dark matter and revolutionizing our understanding of the universe. With ingenious experiments and cutting-edge technology, scientists are pushing the boundaries of knowledge to unravel the secrets hidden within the smallest particles of our cosmos. The search for axions is not just for scientists but for everyone who seeks to comprehend the hidden universe that surrounds us.

Chapter 4: Unraveling the Nature of Dark Matter

The Standard Model and Beyond

In the vast and mysterious realm of particle physics, the Standard Model stands as the cornerstone of our understanding. It is a remarkable framework that describes the fundamental particles and forces that govern our universe. But as we delve deeper into the hidden universe, we realize that the Standard Model is just the beginning of a grander story waiting to be told.

The Standard Model elegantly classifies particles into two categories: quarks and leptons. Quarks are the building blocks of protons and neutrons, while leptons include familiar particles like electrons and neutrinos. These particles interact through four fundamental forces: electromagnetism, the weak force, the strong force, and gravity. The Standard Model beautifully explains how these forces work and how particles interact with each other.

However, the Standard Model is not without its limitations. It fails to account for two perplexing phenomena that dominate the universe: dark matter and dark energy. Dark matter, which outweighs visible matter by a factor of five, is an invisible substance that exerts gravitational pull but does not emit or absorb light. Dark energy, on the other hand, is an enigmatic force that drives the accelerating expansion of the universe. Exploring the mysteries of dark matter and dark energy takes us beyond the confines of the Standard Model.

Scientists are actively searching for evidence of new particles and forces that could extend our knowledge beyond the Standard Model.

They are conducting experiments in massive underground particle colliders and deep space observatories to shed light on the nature of dark matter and dark energy. These endeavors require extraordinary precision and technological advancements.

One promising theory that goes beyond the Standard Model is supersymmetry. It suggests that each particle in the Standard Model has a yet-to-be-discovered superpartner. Supersymmetry could provide a solution to the enigma of dark matter and could help explain why the masses of particles in the universe are so different.

Other theories propose the existence of extra dimensions or new fundamental forces that influence the behavior of particles and energy at the smallest scales. These theories challenge our current understanding and invite us to explore uncharted territories.

As we embark on this journey through the hidden universe, it is important to remember that particle physics is not just the realm of scientists. It is a subject that impacts all of us, as it unravels the mysteries of the universe we inhabit. Understanding the Standard Model and its limitations is crucial for comprehending the fundamental workings of nature and our place in the cosmos.

In the chapters that follow, we will delve deeper into the fascinating world of particle physics, exploring the cutting-edge research and discoveries that pave the way for a new understanding of dark matter, dark energy, and the hidden universe. Join us on this captivating exploration, as we unravel the secrets of the cosmos together.

Supersymmetry and Weakly Interacting Massive Particles (WIMPs)

In the quest to unravel the mysteries of the universe, particle physicists have delved deep into the realm of subatomic particles. This subchapter explores the fascinating concepts of supersymmetry and weakly interacting massive particles (WIMPs) – two crucial elements in our understanding of the hidden universe.

Supersymmetry, often referred to as SUSY, is a theoretical framework that postulates a profound connection between particles with integer spin (bosons) and particles with half-integer spin (fermions). This revolutionary idea proposes that for every known particle in the Standard Model of particle physics, there exists a supersymmetric partner particle. These partner particles have not yet been observed, but their existence is crucial in explaining the properties of dark matter.

Speaking of dark matter, scientists have long been puzzled by its elusive nature. Dark matter is an invisible substance that makes up about 85% of the matter in the universe, yet it does not interact with light or other forms of electromagnetic radiation. This is where weakly interacting massive particles, or WIMPs, come into play. WIMPs are hypothetical particles that are believed to be responsible for the gravitational effects observed in galaxies and galaxy clusters.

Supersymmetry provides a compelling explanation for the existence of WIMPs. According to this theory, the lightest supersymmetric particle (LSP) is a prime candidate for dark matter. The LSP is stable and weakly interacting, making it an ideal candidate for explaining the observed behavior of dark matter. Scientists are actively searching for evidence of WIMPs using a variety of experimental techniques, including underground detectors and particle colliders.

Understanding supersymmetry and WIMPs is crucial not only for particle physicists but also for all of us. These concepts have profound implications for our understanding of the universe and its composition. By unraveling the mysteries of dark matter, we gain deeper insights into the evolution of galaxies, the formation of structures in the universe, and even the fate of our own Milky Way.

In conclusion, supersymmetry and weakly interacting massive particles (WIMPs) offer a tantalizing glimpse into the hidden universe of dark matter. These concepts bridge the gap between the known particles in the Standard Model and the mysterious dark matter that pervades the cosmos. As scientists continue to probe the depths of particle physics, we are edging closer to unraveling the secrets of our universe and gaining a deeper understanding of the forces that shape it.

Other Theoretical Candidates for Dark Matter

In the quest to unravel the mysteries of the universe, scientists have proposed numerous theoretical candidates for dark matter. While the leading candidate, the Weakly Interacting Massive Particle (WIMP), continues to dominate the conversation, there are several other intriguing possibilities worth exploring. In this subchapter, we will delve into some of these alternative theoretical candidates for dark matter, offering a glimpse into the mind-boggling possibilities that lie beyond our current understanding.

One such candidate is the Axion, a hypothetical particle that was first proposed to solve a puzzle in particle physics known as the Strong CP problem. Axions are incredibly light and possess unique properties that make them excellent dark matter candidates. Despite lacking experimental confirmation, scientists are actively searching for axions using various cutting-edge techniques, such as the Axion Dark Matter Experiment (ADMX), which aims to directly detect these elusive particles.

Another fascinating possibility is the Sterile Neutrino, a theoretical particle that does not interact through any of the known forces except gravity. Sterile neutrinos are postulated to have a significant mass and could account for a substantial fraction of dark matter. While their existence is yet to be confirmed, experiments like the IceCube Neutrino Observatory are actively searching for evidence of sterile neutrinos by studying neutrino oscillations and interactions.

Supersymmetric Particles, or sparticles, have long been of interest to particle physicists as potential dark matter candidates. These hypothetical particles arise from supersymmetry, a theory that posits a mirror-like symmetry between fermions and bosons. If supersymmetry is indeed a fundamental aspect of our universe, then the lightest supersymmetric particle (LSP) could be stable and serve as dark matter. Researchers are conducting experiments at the Large Hadron Collider (LHC) to search for evidence of supersymmetry and potentially detect sparticles.

Additionally, the Moduli field, a hypothetical scalar field that emerges from string theory, is another intriguing dark matter candidate. The Moduli field is predicted to have a mass on the order of the Hubble scale and could contribute significantly to the overall dark matter content of the universe. Although experimental evidence for the Moduli field is still lacking, ongoing research in string theory and cosmology continues to shed light on its potential role.

These are just a few examples of the myriad other theoretical candidates for dark matter that exist beyond the realm of WIMPs. While the search for dark matter continues to push the boundaries of human knowledge, the incredible diversity of possibilities highlights the sheer complexity of the universe we inhabit. As scientists strive to unlock the secrets of dark matter, we are reminded that the hidden universe is full of surprises, waiting to be revealed by the relentless pursuit of understanding.

The Role of Particle Colliders in Dark Matter Research

Particle physics, a field that aims to understand the fundamental building blocks of the universe, has made remarkable strides in recent decades. One of the most intriguing mysteries that particle physicists are currently unraveling is the nature of dark matter – the elusive substance that makes up a significant portion of the universe. In this subchapter, we will explore the crucial role that particle colliders play in advancing our understanding of dark matter.

Particle colliders are gigantic machines that accelerate particles to nearly the speed of light and then smash them together. These high-energy collisions allow scientists to probe the fundamental constituents of matter and explore the mysteries of the universe. When it comes to dark matter research, particle colliders serve as powerful tools to investigate the properties and interactions of this enigmatic substance.

Firstly, particle colliders can potentially produce dark matter particles directly. By colliding particles with extremely high energies, scientists hope to generate enough energy to create dark matter particles in the laboratory. These experiments aim to detect the elusive dark matter particles and study their properties, which can provide valuable insights into their nature and behavior.

Secondly, particle colliders can indirectly contribute to dark matter research by studying the known particles that interact with dark matter. Although dark matter does not interact with ordinary matter through electromagnetic forces, it is hypothesized to interact weakly with certain particles, such as neutrinos or the Higgs boson. By colliding particles at high energies, scientists can study the behavior of these known particles and search for any anomalies that may be caused by interactions with dark matter.

Furthermore, particle colliders enable physicists to test various theoretical models that attempt to explain dark matter. These models often involve new particles, forces, or dimensions beyond our current understanding. By simulating these hypothetical scenarios in particle colliders, researchers can investigate the predictions of these models and determine their validity.

Lastly, particle colliders provide a platform for interdisciplinary collaboration. Scientists from different fields, such as astrophysics, cosmology, and particle physics, can come together to exchange ideas and insights. This collaborative effort allows researchers to combine data from various experiments and observations, leading to a more comprehensive understanding of dark matter.

In conclusion, particle colliders play a crucial role in dark matter research by providing the means to directly produce and study dark matter particles, indirectly probe interactions with known particles, test theoretical models, and foster interdisciplinary collaboration. These remarkable machines are at the forefront of scientific exploration, shedding light on the hidden universe of dark matter and paving the way for new discoveries that will revolutionize our understanding of the cosmos.

Chapter 5: The Mysteries of Dark Energy

The Accelerating Expansion of the Universe

In this subchapter, we delve into one of the most fascinating and mind-boggling discoveries of modern science: the accelerating expansion of the universe. This phenomenon has revolutionized our understanding of the cosmos, revealing a hidden universe filled with mysterious dark matter and dark energy.

Our journey begins with the groundbreaking observations made by astronomers in the late 1990s. Up until that point, scientists believed that the expansion of the universe was gradually slowing down due to the gravitational pull of matter. However, to their astonishment, they found evidence that the expansion was actually speeding up.

This discovery challenged everything we thought we knew about the universe's fate. It indicated the existence of a repulsive force, dubbed dark energy, that is driving the acceleration. Dark energy remains one of the greatest mysteries in physics, as its origin and nature are still largely unknown. Yet, its presence is undeniable, shaping the future of our universe.

To understand this concept fully, we must also explore the role of dark matter. While dark energy influences the expansion of the universe on a large scale, dark matter plays a crucial role in the formation and structure of galaxies. Unlike ordinary matter, dark matter does not emit, absorb, or reflect light, making it invisible to our telescopes. However, its presence is inferred through its gravitational effects on visible matter.

Particle physics enters the scene as we explore the nature of dark matter. Scientists have proposed various candidate particles, such as weakly interacting massive particles (WIMPs), that could make up dark matter. Particle accelerators and underground experiments are being used to search for these elusive particles, hoping to shed light on their properties and interactions.

As we peer into the depths of the hidden universe, we come face to face with the mysteries of dark matter and dark energy. Their existence challenges our understanding of the fundamental forces and particles that govern the cosmos. The accelerating expansion of the universe opens up new avenues of research, pushing the boundaries of particle physics and cosmology.

In this subchapter, we aim to unravel the enigmas of the accelerating expansion of the universe. From the observations that led to its discovery to the ongoing quest to understand dark matter and dark energy, our journey will take us on a thrilling exploration of the hidden universe. So buckle up and prepare to embark on a voyage that will redefine our understanding of the cosmos.

The Cosmological Constant

In the vast expanse of the universe, there exists a mysterious force that governs the expansion of space itself. This enigmatic entity is known as the cosmological constant. As we delve into the depths of particle physics, we uncover the intricate workings of this fundamental aspect of our cosmos.

The cosmological constant was first introduced by Albert Einstein in his theory of general relativity. It represents a constant energy density that is inherent to space, driving its expansion. Initially, Einstein introduced this term to counteract the force of gravity, ensuring a static universe. However, it was later discovered that the universe is not static but rather expanding at an accelerating rate.

This revelation led to a reevaluation of the cosmological constant, as it became clear that its role extended beyond countering gravity. Scientists have since postulated that it may be linked to the presence of dark energy, a mysterious force that permeates the cosmos and is responsible for the accelerated expansion. The nature of dark energy and its relationship to the cosmological constant remain one of the most intriguing puzzles of modern physics.

In the realm of particle physics, understanding the cosmological constant holds paramount importance. It provides insights into the fundamental nature of our universe and sheds light on the behavior of dark energy. Researchers have proposed various models and theories to explain the cosmological constant, ranging from modifications of general relativity to the existence of exotic particles.

The cosmological constant also has profound implications for our understanding of the origin and fate of the universe. Its presence and behavior influence the structure and evolution of galaxies, galaxy clusters, and even the distribution of matter on cosmic scales. By unraveling the mysteries of the cosmological constant, we can gain a deeper understanding of the hidden universe that surrounds us.

As we journey through the realms of particle physics, we are constantly confronted with the enigmatic nature of the cosmological constant. It challenges our current understanding of the universe and pushes the boundaries of scientific exploration. The search for answers continues, driven by the curiosity and thirst for knowledge that resides within all of us.

Whether you are an avid enthusiast of particle physics or simply curious about the mysteries of our universe, the cosmological constant offers a captivating glimpse into the hidden workings of the cosmos. Its study invites us to embark on a journey of discovery, where we unravel the secrets of dark matter, dark energy, and the fundamental truths that shape our existence.

Alternative Theories of Dark Energy

In the vast expanse of the cosmos, there exists an invisible force that has puzzled scientists for decades: dark energy. This mysterious entity, responsible for the accelerated expansion of the universe, remains one of the greatest enigmas of modern physics. While the prevailing theory suggests that dark energy is a cosmological constant, there are alternative theories that propose different explanations for this perplexing phenomenon.

One alternative theory posits that dark energy is not a constant, but rather a dynamic field known as quintessence. Unlike the cosmological constant, quintessence fluctuates over time, leading to variations in the expansion rate of the universe. This theory offers a more dynamic and evolving model of dark energy, raising intriguing possibilities for understanding the nature of this elusive force.

Another alternative theory suggests that dark energy is a manifestation of modifications to Einstein's theory of general relativity. In this scenario, gravity behaves differently on cosmic scales, leading to an accelerated expansion of the universe. These modifications could be attributed to additional dimensions or extra fields that influence the behavior of gravity, providing an alternative explanation for the cosmic acceleration.

Furthermore, some physicists propose that dark energy may not exist at all, and the observed accelerated expansion is a result of an incomplete understanding of gravity. This theory challenges the notion of dark energy, arguing that our current understanding of gravity is incomplete and needs refinement. If this theory holds true, it would revolutionize our understanding of the universe and necessitate a reevaluation of fundamental physics principles.

While these alternative theories offer intriguing explanations for dark energy, they are still subject to rigorous scrutiny and testing. Particle physicists around the world are conducting experiments and observations to gather evidence that supports or refutes these alternative theories. By exploring these alternative explanations, scientists hope to gain a deeper understanding of the fundamental forces that shape our universe and unravel the secrets of dark energy.

In conclusion, the study of dark energy continues to captivate the minds of particle physicists and cosmologists alike. The alternative theories presented here offer fresh perspectives on the nature of dark energy, challenging conventional wisdom and pushing the boundaries of our knowledge. Through ongoing research and experimentation, we inch closer to unraveling the mysteries of the hidden universe and shedding light on the enigmatic force that drives the accelerated expansion of our cosmos.

Probing Dark Energy with Supernovae and Cosmic Microwave Background

In the vast expanse of the cosmos, mysteries abound. Among the most enigmatic phenomena are dark matter and dark energy, which together constitute about 95% of our universe. To unravel the secrets of these elusive entities, scientists have turned their attention to the observation of supernovae and the cosmic microwave background.

Supernovae, the explosive deaths of massive stars, offer a unique opportunity to study the expansion history of the universe. By precisely measuring the brightness and redshift of these cosmic beacons, scientists can infer the distances to these supernovae, providing crucial insights into the expansion rate of the universe over time. This groundbreaking work led to the astonishing discovery that the expansion of the universe is accelerating, hinting at the existence of a mysterious force known as dark energy.

But how can we truly understand the nature of dark energy? Enter the cosmic microwave background (CMB), the ancient light left over from the Big Bang. As the universe expanded and cooled, this primordial radiation permeated the cosmos, carrying valuable information about its early stages. By scrutinizing the patterns and fluctuations in the CMB, scientists can decipher the composition of the universe and shed light on the properties of dark energy.

Combining observations of supernovae and the CMB, scientists have embarked on a cosmic journey to understand dark energy. Their investigations have led to the development of intricate theoretical models, including the concept of quintessence, which proposes that dark energy is a dynamic field that evolves over time. Other theories suggest that dark energy may be related to modifications of gravity on cosmic scales.

While the true nature of dark energy remains elusive, these observations have transformed our understanding of the universe. They have challenged the conventional wisdom of a decelerating expansion and opened up new avenues for research in particle physics, cosmology, and fundamental physics.

This subchapter aims to demystify the complexities of probing dark energy with supernovae and the cosmic microwave background. By delving into the intricacies of these cosmic phenomena, it invites readers from all walks of life to embark on a journey through the hidden universe. Whether you are a curious enthusiast or a seasoned particle physicist, this exploration will deepen your appreciation for the wonders of our cosmos and the ongoing quest to unravel its secrets. Join us on this captivating odyssey and discover the hidden universe of dark matter and dark energy.

Chapter 6: The Cosmic Web: Dark Matter's Influence on Structure Formation

The Birth of the Universe

In the vast expanse of the cosmos, a remarkable event occurred billions of years ago - the birth of the universe. This awe-inspiring moment, often referred to as the Big Bang, marked the beginning of everything we know today. Exploring the origins of the universe requires delving into the realm of particle physics, a captivating field that unravels the mysteries of matter and energy.

At its core, particle physics seeks to understand the fundamental building blocks of the universe and the forces that govern them. By studying the tiniest particles known to humanity, such as quarks and leptons, scientists can piece together the puzzle of our existence. The birth of the universe is intricately connected to these particles and the immense energy that was unleashed during the Big Bang.

According to the prevailing theory, the universe sprang into existence from an unimaginably dense and hot state. At this moment, all matter and energy were compressed into a singularity - a point of infinite density. As the universe rapidly expanded, particles and antiparticles were created and annihilated in a continuous dance. However, a slight asymmetry between matter and antimatter allowed some particles to survive this annihilation, leading to the formation of the matter-dominated universe we now inhabit.

As the universe cooled, particles began to come together, forming protons and neutrons. These particles, in turn, combined to create the nuclei of light elements like hydrogen and helium. Over billions of years, gravity took hold, causing these elements to clump together and form stars and galaxies. These cosmic structures became the birthplaces of new elements, generated through the nuclear reactions that occur within the cores of stars.

The birth of the universe also gave rise to another enigmatic phenomenon - dark matter and dark energy. While these entities cannot be directly observed, their presence can be inferred through their gravitational effects on visible matter. Dark matter, which comprises around 27% of the universe, exerts a gravitational pull on galaxies, holding them together. Dark energy, on the other hand, is responsible for the accelerated expansion of the universe, pushing galaxies apart.

Understanding the birth of the universe is an ongoing quest that pushes the boundaries of human knowledge. Particle physics provides a window into the remarkable events that unfolded during the Big Bang and continues to shape our understanding of the cosmos. By delving into the mysteries of the universe's origins, we gain insights into the fundamental nature of reality and our place within it.

Gravitational Clustering

In the vast expanse of the universe, where countless galaxies reside, a fascinating phenomenon known as gravitational clustering takes place. This captivating process plays a crucial role in shaping the structure of the cosmos as we know it. Let us embark on a journey to explore the intricacies of this phenomenon and its significance in the realm of particle physics.

Gravitational clustering refers to the tendency of matter, both visible and invisible, to group together under the influence of gravity. It is a delicate dance between the attractive force of gravity and the initial density fluctuations present in the early universe. These tiny variations in the density of matter, imprinted during the cosmic inflation period, set the stage for the formation of galaxies, galaxy clusters, and even larger cosmic structures.

As matter begins to clump together, the gravitational force intensifies, drawing more nearby matter towards these regions of higher density. Over time, these clumps grow, attracting more and more matter, creating a web-like structure called the cosmic web. This intricate network of filaments and voids is the result of gravitational clustering, weaving together the fabric of the universe on a grand scale.

Gravitational clustering is not limited to visible matter alone. Dark matter, an enigmatic substance that outweighs visible matter by a factor of five, plays a significant role in this process. Although its nature remains elusive, its gravitational influence is undeniable. Dark matter acts as the scaffolding on which galaxies and galaxy clusters form, providing the gravitational glue necessary for the clumping of visible matter.

Particle physicists are particularly intrigued by gravitational clustering as it offers a unique window into the nature of dark matter and dark energy. By studying the distribution and behavior of matter on both small and large scales, scientists can gain valuable insights into the fundamental particles that make up our universe.

Through sophisticated observations and computer simulations, researchers can trace the evolution of cosmic structures, revealing the intricate interplay between gravity, dark matter, and visible matter. These studies not only enhance our understanding of the universe's past but also shed light on its future trajectory.

Gravitational clustering is a testament to the remarkable beauty and complexity of our universe. It is a phenomenon that shapes galaxies, drives cosmic evolution, and holds the keys to unraveling the mysteries of particle physics. By delving deeper into this captivating process, we take another step towards unraveling the hidden universe and our place within it.

Simulations and Observations of Large-Scale Structure

In the vast expanse of the universe, a hidden structure exists that shapes the very fabric of our cosmos. This structure, known as the large-scale structure, is a web-like network of galaxies, galaxy clusters, and cosmic filaments that spans billions of light-years. It is through simulations and observations that we have come to understand the intricacies of this hidden universe.

Particle physics plays a crucial role in unraveling the mysteries of the large-scale structure. By studying the fundamental particles and their interactions, scientists can gain insights into the formation and evolution of galaxies and galaxy clusters. Through computer simulations, researchers can recreate the conditions of the early universe and observe how these particles interacted to give rise to the structures we see today.

One of the most remarkable tools in this quest is the use of supercomputers to simulate the growth of cosmic structures. These simulations involve complex algorithms that take into account the laws of physics and the initial conditions of the early universe. By following the evolution of billions of particles over billions of years, scientists can recreate the large-scale structure with remarkable accuracy.

These simulations have revealed that the distribution of matter in the universe is far from random. Instead, it forms a web-like pattern, with dense regions known as galaxy clusters connected by vast cosmic filaments. These filaments act as highways of matter, allowing galaxies to form and evolve along their length. The simulations also show that dark matter, an elusive substance that does not emit, absorb, or reflect light, plays a crucial role in shaping the large-scale structure.

Observations of the large-scale structure further validate these simulations. Telescopes, both on the ground and in space, have provided stunning images of the cosmic web, revealing the intricate patterns and clustering of galaxies. By studying the distribution and motion of galaxies, astronomers can measure the expansion rate of the universe and infer the presence of dark matter and dark energy.

Understanding the large-scale structure is not just an academic pursuit; it has profound implications for our understanding of the universe as a whole. By studying the formation and growth of galaxies and galaxy clusters, scientists can uncover the secrets of dark matter and dark energy, two enigmatic components that make up the majority of the universe's composition. Furthermore, these findings can help us comprehend the origins of our own cosmic neighborhood and shed light on the fate of the universe itself.

In conclusion, simulations and observations of the large-scale structure provide a window into the hidden universe of dark matter and dark energy. Through the lens of particle physics, we can unravel the mysteries of the cosmos and gain a deeper understanding of our place within it. By exploring this fascinating subchapter, we embark on a journey that bridges the realms of particle physics and the wonders of the universe, inviting everyone to join in the exploration of the hidden universe.

Dark Matter Halos and Galaxy Formation

In the vast expanse of the universe, there lies a mysterious substance that has captivated the minds of scientists for decades - dark matter. This enigmatic entity makes up about 85% of the total matter in the cosmos, yet its true nature remains elusive. As we delve into the depths of particle physics, we begin to unravel the secrets of dark matter halos and their profound influence on galaxy formation.

Dark matter halos are vast, invisible structures that span across galaxies. They are composed of dark matter particles, which do not interact with light or other electromagnetic radiation. Instead, their presence is inferred through their gravitational effects on visible matter. Like an invisible puppeteer, dark matter shapes the cosmic web, guiding the formation of galaxies and galaxy clusters.

The formation of galaxies begins with tiny fluctuations in the density of dark matter during the early universe. These fluctuations serve as the seeds for the growth of structures. Over billions of years, gravity causes dark matter to clump together, forming halos of various sizes. These halos act as cosmic nurseries, attracting and capturing ordinary matter, such as gas and dust.

As matter accumulates within these halos, it collapses under its own gravity, forming dense regions known as protogalactic clouds. Within these clouds, stars are born, bringing light and life to the otherwise dark and mysterious universe. The gravitational pull of dark matter halos also influences the motion of stars within galaxies, ensuring their stability and coherence.

Understanding the properties of dark matter and its halos is crucial to comprehending the evolution of galaxies and the cosmos itself. Through meticulous observations and sophisticated computer simulations, scientists have been able to map the distribution of dark matter halos across the universe and trace their impact on the formation and evolution of galaxies.

The study of dark matter halos and galaxy formation not only sheds light on the hidden universe but also holds implications for our understanding of fundamental physics. The nature of dark matter remains one of the biggest unsolved puzzles in modern science, and unraveling its mysteries could revolutionize our understanding of the fundamental building blocks of the universe.

In conclusion, dark matter halos are the invisible scaffolding that shape the cosmos. They provide the gravitational framework for the formation and evolution of galaxies, fostering the birth of stars and the creation of life. Exploring the depths of particle physics is not only a journey through the hidden universe but also a quest to unravel the enigma of dark matter and its profound influence on the cosmos.

Chapter 7: The Future of Dark Matter and Dark Energy Research

Advancements in Experimental Techniques

Particle Physics has always pushed the boundaries of scientific exploration, delving into the mysteries of the universe at the tiniest scales. Over the years, experimental techniques in this field have witnessed remarkable advancements, allowing scientists to delve deeper into the hidden universe of dark matter and dark energy. These breakthroughs have not only expanded our understanding of the cosmos but have also paved the way for numerous technological advancements that benefit us all.

One of the most significant advancements in experimental techniques is the development of particle accelerators. These mammoth machines propel particles at incredible speeds, allowing scientists to recreate the conditions of the early universe. With each new generation of accelerators, we gain a deeper insight into the fundamental building blocks of matter and the forces that govern them. From the Large Hadron Collider (LHC) to the proposed Future Circular Collider (FCC), these colossal machines have revolutionized our understanding of particle physics.

Another key advancement is the development of sophisticated detectors. These devices capture the elusive particles produced in high-energy collisions, providing valuable data for physicists to analyze. Detectors like the ATLAS and CMS at the LHC have played a crucial role in the discovery of the Higgs boson, a groundbreaking achievement that confirmed the existence of the Higgs field, which gives particles mass. Furthermore, advancements in detector technology have enabled scientists to study neutrinos, elusive particles that can provide insights into the nature of dark matter and dark energy.

Moreover, advancements in imaging techniques have enhanced our ability to observe the invisible. With the advent of advanced telescopes and satellites, scientists have been able to map the distribution of dark matter and detect its gravitational effects on visible matter. These observations have not only confirmed the existence of dark matter but have also shed light on its role in shaping the structure of galaxies and the universe as a whole.

The progress made in experimental techniques has not only benefited the field of particle physics but has also had a significant impact on various other scientific disciplines. The technologies developed for particle accelerators and detectors have found applications in medicine, materials science, and even environmental studies. From cancer treatments to high-performance materials, these advancements have improved the quality of life for people from all walks of life.

In conclusion, advancements in experimental techniques have propelled the field of particle physics forward, unveiling the secrets of dark matter and dark energy. These breakthroughs have not only expanded our knowledge of the universe but have also led to numerous technological advancements that benefit us all. As our journey through the hidden universe continues, we can expect even more remarkable discoveries and innovations in the years to come.

The Role of Space Missions

Space missions play a vital role in our understanding of the hidden universe, providing us with invaluable insights into the mysteries of dark matter and dark energy. These ambitious endeavors allow us to explore the cosmos beyond our planet, providing a unique vantage point that complements the observations made from Earth.

In the realm of particle physics, space missions are particularly significant. They enable us to study high-energy cosmic rays and particles that are not easily detectable from the ground. By launching sophisticated instruments into space, scientists can capture and analyze particles that hold the key to unlocking the secrets of the universe.

One of the primary objectives of space missions is to search for dark matter. This elusive substance, which constitutes about 27% of the universe, neither emits nor reflects light, making it nearly impossible to observe directly. However, through space missions, we have been able to indirectly detect the presence of dark matter by observing its gravitational effects on visible matter. By mapping the distribution of dark matter in the universe, we can gain valuable insights into its nature and properties.

Additionally, space missions allow us to study dark energy, another enigmatic component of the universe. Dark energy is believed to be responsible for the accelerated expansion of the universe, but its origin and composition remain largely unknown. By using space-based instruments to observe distant supernovae and measure cosmic microwave background radiation, scientists can gather data to better understand this mysterious force.

Moreover, space missions provide a unique opportunity to observe celestial bodies and phenomena that are obscured or distorted by Earth's atmosphere. By escaping the limitations of atmospheric interference, scientists can obtain clearer and more precise measurements, enhancing our understanding of the universe's fundamental particles and their interactions.

Furthermore, space missions facilitate international collaboration, bringing together scientists from different countries and backgrounds to work towards a common goal. These collaborative efforts not only help to advance our knowledge but also foster goodwill and cooperation among nations.

In conclusion, space missions are indispensable tools in the field of particle physics. They allow us to explore the hidden universe, study dark matter and dark energy, and observe celestial objects with unprecedented clarity. By pushing the boundaries of human knowledge, space missions contribute to our understanding of the universe and inspire future generations to continue unraveling its mysteries.

Multi-Messenger Astronomy and Dark Matter

In the vast expanse of the universe, there are hidden mysteries waiting to be unraveled. These enigmas lie in the realms of dark matter and dark energy, which have baffled scientists for decades. However, recent advancements in the field of multi-messenger astronomy have shed new light on these elusive entities, bringing us closer to understanding the hidden universe.

Multi-messenger astronomy is a revolutionary approach that combines data from various cosmic messengers, such as light, neutrinos, and gravitational waves, to paint a more comprehensive picture of the cosmos. This interdisciplinary field has opened up new avenues for research, allowing scientists to explore the fundamental nature of our universe in unprecedented ways.

One of the most intriguing aspects of multi-messenger astronomy is its role in studying dark matter. Dark matter is an invisible substance that makes up a significant portion of the universe, exerting its gravitational pull on the visible matter we can see. Despite its dominance, dark matter remains elusive, as it does not emit, absorb, or reflect any form of electromagnetic radiation. This is where multi-messenger astronomy comes into play.

By analyzing a combination of different cosmic messengers, scientists can indirectly detect the presence of dark matter. For instance, gravitational lensing, which occurs when the gravitational pull of dark matter bends the path of light, can provide valuable clues about its distribution in the universe. Moreover, the detection of high-energy neutrinos, which are produced during violent cosmic events like supernovae, can help map the distribution of dark matter in galaxies and galaxy clusters.

The exploration of dark matter through multi-messenger astronomy not only deepens our understanding of the universe but also has profound implications for particle physics. Particle physicists are constantly searching for new particles and interactions that can explain the mysteries of the universe. Dark matter, being a crucial missing piece of the cosmic puzzle, is believed to be composed of exotic particles that interact weakly with ordinary matter.

By combining the insights from multi-messenger astronomy with particle physics experiments, scientists hope to unveil the true identity of dark matter. This collaboration between the two fields has the potential to revolutionize our understanding of the universe, revealing new particles, interactions, and even dimensions that have so far remained hidden from our view.

In conclusion, the emergence of multi-messenger astronomy has brought us closer to unraveling the secrets of the hidden universe. By combining data from various cosmic messengers, scientists can probe the mysteries of dark matter and its implications for particle physics. This interdisciplinary approach holds the promise of transforming our understanding of the universe, taking us on a fascinating journey through the realms of dark matter and dark energy.

The Quest for a Unified Theory of Everything

In the fascinating world of particle physics, scientists are embarking on a grand adventure - the quest for a unified theory of everything. This ambitious pursuit aims to explain the fundamental forces and particles that govern our universe, bringing together the realms of the very small and the very large. Welcome to a subchapter of "The Hidden Universe: A Journey through Dark Matter and Dark Energy for Everyone," where we delve into the mysteries of this quest and its profound implications.

At the heart of the exploration lies the desire to reconcile two seemingly incompatible theories: general relativity and quantum mechanics. General relativity describes the behavior of gravity on cosmic scales, while quantum mechanics unravels the behavior of particles on the subatomic level. Yet, when combined, these theories clash, leaving physicists yearning for a more comprehensive framework that unifies all known phenomena.

The quest begins with understanding the fundamental particles that make up matter. Scientists have discovered an intricate web of particles, each with unique properties and behaviors. From the elegant simplicity of quarks to the elusive nature of neutrinos, these building blocks of the universe hold the key to unlocking its deepest secrets.

Enter the concept of supersymmetry, an intriguing theory suggesting that every known particle has a "superpartner." These superpartners, if found, could help bridge the gap between general relativity and quantum mechanics, offering a glimpse into a unified theory. Researchers around the world are tirelessly searching for evidence of these elusive counterparts, utilizing massive particle accelerators and innovative detection techniques.

The quest for a unified theory of everything also takes us beyond the known particles, into the realm of dark matter and dark energy. These mysterious entities, which make up the majority of the universe, exert a gravitational pull but remain invisible to our traditional methods of detection. Unraveling the nature of dark matter and dark energy is vital to understanding the cosmic web in which we reside.

As we embark on this journey, it is important to remember that the quest for a unified theory of everything is not confined to the realm of academia. It has far-reaching implications for our everyday lives. The technologies that emerge from these explorations could revolutionize fields such as energy, computing, and medicine. By unraveling the secrets of the universe, we may unlock the key to our own advancement as a species.

So, whether you are a curious individual, a lover of physics, or someone seeking to grasp the mysteries that lie beyond the visible, join us on this awe-inspiring journey. The quest for a unified theory of everything is a testament to human curiosity and our relentless pursuit of knowledge. Together, let us unravel the hidden universe and illuminate the path to a deeper understanding of our existence.

Chapter 8: Implications and Applications of Dark Matter and Dark Energy

Dark Matter in the Milky Way and Beyond

As we gaze up at the night sky, we are captivated by the beauty and mystery of the Milky Way. But what lies beyond the twinkling stars and swirling galaxies? The answer is dark matter, a mysterious substance that pervades our universe, yet remains largely elusive to our understanding. In this subchapter, we will embark on a journey through the hidden universe of dark matter, exploring its presence in the Milky Way and beyond.

To comprehend the enigma of dark matter, we must delve into the realm of particle physics. At its core, particle physics seeks to understand the fundamental building blocks of our universe. It is within this framework that we encounter the tantalizing mysteries surrounding dark matter. Scientists believe that dark matter consists of particles that do not interact with light, making it invisible to our telescopes and detectors. This poses a significant challenge in our attempts to study and unravel its secrets.

Within the Milky Way, dark matter plays a pivotal role in sculpting the structure of our galaxy. Through its gravitational influence, dark matter acts as an invisible scaffolding, guiding the formation and movement of stars and galaxies. Without dark matter, the Milky Way as we know it would not exist. Yet, despite its undeniable impact, we are still grappling to comprehend its true nature.

Beyond our galaxy, dark matter extends its reach across the cosmos, binding together vast clusters of galaxies and shaping the cosmic web. Its influence is felt on the largest scales, holding the universe in its mysterious embrace. The quest to understand the role of dark matter in the wider universe has led scientists to develop ingenious experiments and observatories, pushing the boundaries of our knowledge.

While dark matter continues to elude direct detection, scientists have made significant progress in uncovering its fingerprints. By studying the motions of stars and galaxies, as well as the bending of light by gravitational lensing, we have amassed compelling evidence for the existence of dark matter. These observations have opened up new avenues of research, deepening our understanding of the hidden universe that lies beyond our senses.

In this subchapter, we have barely scratched the surface of the captivating journey through dark matter in the Milky Way and beyond. The mysteries surrounding this elusive substance continue to intrigue and inspire scientists and enthusiasts alike. As we delve deeper into the hidden universe of dark matter, we come face to face with the profound questions that drive particle physics and the tantalizing possibility of unraveling the secrets of our cosmos.

Dark Energy's Influence on the Fate of the Universe

In the vast expanse of the cosmos, there exists an enigmatic force that holds the key to understanding the fate of our universe. This force, known as dark energy, has captivated the minds of scientists and astronomers alike. In the subchapter titled "Dark Energy's Influence on the Fate of the Universe," we delve into the profound implications this mysterious entity has on the fundamental nature of our existence.

For the audience of "everyone," this chapter aims to make the complex field of particle physics accessible and comprehensible to all. We embark on a journey through the hidden universe of dark matter and dark energy, shedding light on the intricate workings of the cosmos.

Dark energy, a repulsive force that counteracts gravity, plays a pivotal role in shaping the future of our universe. It is believed to be responsible for the accelerated expansion of space itself, pushing galaxies apart at an ever-increasing rate. This revelation, initially met with skepticism, has revolutionized our understanding of the cosmos and its ultimate destiny.

Drawing on the field of particle physics, we explore the various theories and hypotheses surrounding dark energy. From the notion of a cosmological constant, proposed by Albert Einstein himself, to more recent theories involving quantum vacuum fluctuations, each concept offers a unique perspective on the nature and origin of this elusive force.

One of the profound questions we address is whether dark energy is a constant presence throughout the universe's history or if it evolves over time. Scientists are actively investigating whether dark energy's influence fluctuates, potentially leading to the collapse or eternal expansion of the universe. We delve into these theories, examining the evidence and implications for the fate of our cosmos.

Moreover, we discuss the interconnectedness of dark energy, dark matter, and ordinary matter. While dark matter provides the gravitational scaffolding to hold galaxies together, dark energy seemingly drives them apart. Understanding this delicate interplay between the visible and invisible components of the universe is crucial in unraveling the mysteries of our existence.

Throughout this subchapter, we strive to inspire curiosity and ignite a sense of wonder about the hidden universe that surrounds us. By exploring the influence of dark energy on the fate of the universe, we hope to instill a deeper appreciation for the incredible complexity and beauty of the cosmos we call home. Whether you are a novice or an enthusiast of particle physics, "The Hidden Universe: A Journey through Dark Matter and Dark Energy for Everyone" invites you to embark on a remarkable expedition through the mysteries of our universe.

Implications for Astrophysics and Cosmology

The field of astrophysics and cosmology has been revolutionized by the discovery and study of dark matter and dark energy. These enigmatic components of the universe have profound implications for our understanding of the cosmos, and their exploration has opened up new avenues of research in particle physics.

One of the key implications of dark matter is its role in the formation and evolution of galaxies. Observations have revealed that galaxies are embedded in massive halos of dark matter, which provide the gravitational scaffolding for the formation of stars and galaxies. Without dark matter, the universe as we know it would look drastically different. Understanding the properties and behavior of dark matter is therefore crucial in unveiling the mysteries of galaxy formation.

Furthermore, the existence of dark matter has implications for the search for new particles in particle physics. Many theories beyond the Standard Model predict the existence of new particles that could constitute dark matter. By studying the properties of dark matter, particle physicists hope to shed light on the fundamental nature of matter and the forces that govern the universe. The discovery of dark matter particles would be a major breakthrough in our understanding of the fundamental building blocks of the universe.

On the other hand, dark energy presents a different set of implications for astrophysics and cosmology. Dark energy is responsible for the accelerated expansion of the universe, a discovery that earned the Nobel Prize in Physics in 2011. Understanding the nature of dark energy is one of the greatest challenges in modern science. Its presence suggests that our current understanding of the fundamental forces of nature is incomplete.

Moreover, the existence of dark energy raises questions about the ultimate fate of the universe. Will the accelerated expansion continue indefinitely, leading to a "Big Freeze" scenario where galaxies move further apart and eventually fade away? Or will the expansion halt and reverse, resulting in a "Big Crunch" where the universe collapses in on itself? These questions drive researchers to delve deeper into the nature of dark energy and its implications for the future of our universe.

In conclusion, the study of dark matter and dark energy has revolutionized the field of astrophysics and cosmology. From providing the gravitational scaffolding for galaxy formation to challenging our understanding of the fundamental forces of nature, these enigmatic entities have opened up new avenues of research in particle physics. Exploring the implications of dark matter and dark energy is not only crucial for scientists in these fields but also for anyone interested in unraveling the mysteries of the hidden universe.

The Potential for Technological Breakthroughs

In the ever-evolving landscape of science and technology, the potential for groundbreaking discoveries and technological breakthroughs seems limitless. This is particularly true in the field of particle physics, where scientists continuously push the boundaries of our understanding of the universe. As we embark on a journey through dark matter and dark energy, it is crucial to explore the immense potential that lies within this realm and the transformative impact it can have on our lives.

Particle physics, often referred to as the fundamental science, seeks to unravel the mysteries of the universe by studying the smallest building blocks of matter and the forces that govern them. It is a field that constantly challenges our preconceived notions and drives us towards innovation and new frontiers. With every breakthrough in particle physics, there is a ripple effect that extends far beyond the laboratory walls, influencing various technological advancements that shape our daily lives.

One area where particle physics has already made significant contributions is in medical imaging. The development of positron emission tomography (PET) scanners, for instance, stemmed from the fundamental understanding of particle interactions. These scanners have revolutionized the diagnosis and treatment of diseases, allowing doctors to visualize and monitor the metabolic processes in the human body. This is just one example of how particle physics has directly impacted healthcare, providing doctors with powerful tools for early detection and precise treatment options.

Another exciting area of potential lies in energy production and storage. As we grapple with the challenges of climate change and the need for sustainable energy sources, particle physics offers promising solutions. Advancements in fusion energy, inspired by the study of plasma physics, hold the key to clean and virtually limitless energy generation. Scientists are working towards harnessing the power of the Sun, replicating the fusion reactions that fuel our star. If successful, this could revolutionize our energy infrastructure and help combat climate change.

Furthermore, particle physics research has the potential to drive advancements in computing and data storage. As scientists delve deeper into the mysteries of the universe, they generate vast amounts of data that require cutting-edge technologies for processing and storage. This demand has spurred the development of faster and more efficient computing systems, paving the way for advancements such as quantum computing. Quantum computers, with their ability to process vast amounts of information simultaneously, have the potential to revolutionize fields like cryptography, optimization, and drug discovery.

In conclusion, the potential for technological breakthroughs in particle physics is immense. From healthcare to energy production and computing, the fundamental understanding of particles and their interactions has the power to transform our world. As we embark on this journey through dark matter and dark energy, we must not only appreciate the advancements made in our quest for knowledge but also recognize the profound impact they can have on our daily lives. By investing in particle physics research and fostering collaboration between scientists and engineers, we can unlock the hidden potential of the universe and usher in a new era of technological innovation for the benefit of everyone.

Chapter 9: The Human Connection: Why Dark Matter and Dark Energy Matter to Everyone

The Quest for Knowledge and Understanding

In the vast expanse of the cosmos, there lies a hidden universe, one that defies our conventional understanding of matter and energy. This hidden universe is shrouded in mystery, waiting to be unravelled by those who dare to embark on a journey of knowledge and understanding. Welcome to "The Hidden Universe: A Journey through Dark Matter and Dark Energy for Everyone."

This subchapter delves into the captivating quest for knowledge and understanding in the realm of particle physics. Particle physics, often referred to as the building blocks of the universe, unravels the secrets of the fundamental particles that make up everything we see around us. From the smallest subatomic particles to the grandeur of the cosmos, particle physics offers a window into the mysteries of our existence.

The pursuit of knowledge in particle physics is a story of human curiosity, perseverance, and ingenuity. It is a journey that has spanned centuries, tracing its roots to ancient civilizations' attempts to comprehend the nature of matter and energy. From the groundbreaking experiments of Galileo and Newton to the revolutionary discoveries of Einstein and Bohr, each step forward has brought us closer to unlocking the hidden universe.

Through this subchapter, we invite everyone to join us on this exhilarating expedition. Whether you are a seasoned physicist or simply someone with a curious mind, this journey is for you. We will explore the fundamental particles that make up everything in the universe and how they interact with one another. From quarks to leptons, we will delve into the intricate web of particles that form the tapestry of our reality.

But our quest does not stop there. We will also venture into the realms of dark matter and dark energy, the enigmatic forces that dominate the universe. These invisible entities hold the key to understanding the structure and evolution of galaxies, yet their nature remains elusive. We will unravel the mysteries surrounding dark matter and dark energy, discussing the leading theories and ongoing experiments that strive to shed light on this hidden universe.

"The Hidden Universe: A Journey through Dark Matter and Dark Energy for Everyone" is an invitation to explore the wonders of particle physics and the profound questions it seeks to answer. It is an opportunity to embark on a quest for knowledge and understanding, to uncover the secrets of the hidden universe and glimpse the beauty and complexity that lies within. So, come and join us on this extraordinary journey, where science and imagination collide, and let us embark on an adventure that will forever change the way we perceive the cosmos.

Dark Matter and Dark Energy in Popular Culture

In recent years, the concepts of dark matter and dark energy have captured the imagination of scientists and the general public alike. These mysterious entities, which make up the vast majority of our universe, have also found their way into popular culture. From movies and TV shows to books and video games, dark matter and dark energy have become key elements in storytelling, captivating audiences with their enigmatic nature and potential implications.

One of the most notable examples of dark matter in popular culture is the Marvel Cinematic Universe. In the movie "Thor: The Dark World," the Aether, a powerful substance that is said to be one of the Infinity Stones, is described as being made of dark matter. This portrayal of dark matter as a potent and otherworldly force adds an intriguing layer to the story, enhancing the sense of mystery and wonder surrounding the Aether.

In the realm of literature, dark matter and dark energy often serve as plot devices. Authors have used these concepts to create intricate and thought-provoking narratives. In his novel "Dark Matter," Blake Crouch explores the idea of parallel universes and the influence of dark matter on human existence. The book takes readers on a thrilling journey through different realities, challenging their perception of reality and the role of dark matter within it.

Video games have also embraced the allure of dark matter and dark energy. In the popular Mass Effect series, dark energy is depicted as a potentially catastrophic force that threatens the existence of the universe. Players must navigate through a complex storyline, making choices that determine the fate of galaxies and the balance between dark energy and conventional matter.

While these portrayals in popular culture often take creative liberties with the science behind dark matter and dark energy, they have undoubtedly sparked curiosity and interest in these fascinating phenomena. For the general audience, these depictions provide a gateway to learning more about particle physics and the cutting-edge research being conducted in the field.

"The Hidden Universe: A Journey through Dark Matter and Dark Energy for Everyone" aims to bridge the gap between popular culture and particle physics, offering readers an accessible and engaging exploration of the subject. By examining the way dark matter and dark energy are portrayed in movies, books, and video games, the book invites readers to delve deeper into the scientific concepts behind these phenomena. It not only entertains but also educates, fostering a greater appreciation for the wonders of our universe and the ongoing quest to unravel its mysteries.

Whether you are a casual fan of science fiction or a dedicated particle physics enthusiast, this subchapter will take you on a captivating journey through the influence of dark matter and dark energy in popular culture. Prepare to be amazed and inspired as you discover how these enigmatic entities continue to shape our imagination and our understanding of the hidden universe.

Inspiring the Next Generation of Scientists

Science has always been a fascinating realm, unlocking the mysteries of the universe and expanding our understanding of the world we inhabit. However, in recent years, there has been a growing concern about the declining interest in scientific fields, especially among the younger generation. As we stand at the precipice of a new era, where discoveries in particle physics are reshaping our understanding of the universe, it is crucial to inspire and engage the next generation of scientists.

"The Hidden Universe: A Journey through Dark Matter and Dark Energy for Everyone" aims to bridge the gap between the complex world of particle physics and the general audience. This subchapter, "Inspiring the Next Generation of Scientists," is dedicated to igniting curiosity and fueling the passion of young minds towards scientific exploration.

Science education should not be limited to textbooks and classrooms alone. It is essential to provide young individuals with hands-on experiences that allow them to explore and experiment. Encouraging participation in science fairs, science clubs, and workshops can foster a sense of wonder and excitement about the field. By showcasing the real-world applications of scientific theories and experiments, we can inspire the next generation to pursue careers in particle physics and related disciplines.

Furthermore, it is crucial to highlight the diverse and inclusive nature of scientific pursuits. By introducing young individuals to the stories and achievements of scientists from different backgrounds and cultures, we can break down barriers and encourage a more diverse representation in the field. This subchapter will explore the accomplishments of renowned physicists and their contributions to particle physics, serving as inspiration for aspiring scientists from all walks of life.

Additionally, the subchapter will delve into the advancements and breakthroughs in particle physics that have the potential to revolutionize our understanding of the universe. From the discovery of the Higgs boson to the search for dark matter and dark energy, readers will be introduced to the mysteries that lie at the forefront of scientific exploration. By showcasing the real-life implications of these discoveries, we aim to kindle a sense of purpose and urgency within young minds.

Ultimately, "Inspiring the Next Generation of Scientists" serves as a call to action, inviting individuals from all backgrounds to embark on a journey of scientific discovery. By nurturing the curiosity and passion of young minds, we can ensure a bright future for particle physics and contribute to the collective endeavor of unraveling the hidden secrets of the universe.

The Role of Citizen Science in Dark Matter and Dark Energy Research

In the fascinating world of particle physics, there are still many mysteries waiting to be unraveled. Two of the most enigmatic concepts that scientists are exploring are dark matter and dark energy. These elusive entities make up a significant portion of our universe, yet we know very little about them. However, thanks to the power of citizen science, the field of dark matter and dark energy research has made significant strides.

Citizen science is a collaborative approach that involves members of the general public in scientific research. It provides an opportunity for enthusiasts from all walks of life to contribute to the advancement of knowledge. In the realm of particle physics, citizen scientists have a crucial role to play. Their participation not only expands the scope of research but also accelerates the progress in understanding the hidden universe.

One of the ways citizen scientists contribute to dark matter and dark energy research is through data analysis. Large-scale experiments, such as the Large Hadron Collider, generate an enormous amount of data that can be overwhelming for researchers to analyze on their own. Citizen scientists can assist by sifting through this data, identifying patterns, and making valuable observations. Their collective efforts help scientists identify potential signals or anomalies that could lead to groundbreaking discoveries.

Another significant contribution citizen scientists make is in the realm of creativity and innovation. These enthusiasts come from diverse backgrounds, bringing fresh perspectives and unconventional ideas to the table. Their unique insights often challenge traditional scientific thinking and open new avenues of exploration. By encouraging outside-the-box thinking, citizen science promotes a more holistic approach to dark matter and dark energy research.

Moreover, citizen science initiatives foster a sense of community and engagement among participants. By involving the public in scientific endeavors, citizens gain a deeper understanding of the scientific process and the complexities of particle physics. This engagement not only empowers individuals but also creates a bridge between scientists and the wider society. It leads to increased awareness, appreciation, and support for scientific research.

In conclusion, citizen science has emerged as a powerful tool in the quest to unravel the mysteries of dark matter and dark energy. By harnessing the collective intelligence and creativity of individuals from all walks of life, researchers in the field of particle physics can accelerate their progress. Citizen scientists play a vital role in data analysis, creativity, and community engagement. Through their contributions, they bring us closer to understanding the hidden universe that surrounds us all.

Chapter 10: Conclusion

Recap of Key Discoveries and Concepts

In this subchapter, we will recap some of the key discoveries and concepts that we have explored so far in our journey through the hidden universe of dark matter and dark energy. This recap is especially important for those interested in the field of particle physics, as it will help solidify your understanding of the intricate nature of our universe.

One of the fundamental concepts we encountered is the existence of dark matter, an invisible and mysterious substance that makes up about 27% of the universe. Through various observations and calculations, scientists have deduced its presence by studying the gravitational effects it has on visible matter. Dark matter plays a crucial role in the formation and evolution of galaxies, acting as a scaffolding upon which visible matter clumps and structures itself.

Another concept we explored is dark energy, an even more enigmatic force that makes up approximately 68% of the universe. Dark energy is responsible for the accelerated expansion of our universe, countering the gravitational pull of matter and pushing galaxies away from each other. While its exact nature remains elusive, we discussed some of the leading theories and hypotheses that attempt to explain this enigmatic force.

We delved into the fascinating world of particle physics, the branch of science that studies the fundamental particles and forces of nature. We explored the Standard Model, a theoretical framework that elegantly explains the interactions between particles and the four fundamental forces: electromagnetic, weak, strong, and gravitational (although gravity has yet to be fully incorporated into the model). We discussed the discovery of the Higgs boson, a particle that gives mass to other particles and provides further evidence for the validity of the Standard Model.

Furthermore, we examined the Large Hadron Collider (LHC), the most powerful particle accelerator ever built. The LHC allows scientists to recreate the conditions present shortly after the Big Bang, enabling them to study particles at energies never before achieved. We highlighted some of the groundbreaking discoveries made at the LHC, such as the confirmation of the Higgs boson's existence.

To conclude this recap, we emphasized the importance of ongoing research and the pursuit of knowledge about dark matter and dark energy. These fields continue to evolve, with new discoveries and breakthroughs occurring regularly. By staying informed and engaging with the scientific community, we can all contribute to unraveling the mysteries of the hidden universe and understanding the fundamental nature of our existence.

Whether you are an avid enthusiast or just beginning to explore the world of particle physics, this recap serves as a reminder of the remarkable discoveries and concepts that have shaped our understanding of the hidden universe. It is an invitation to continue your journey into the depths of dark matter and dark energy, where countless mysteries await and new knowledge beckons.

Looking Ahead: The Continued Exploration of the Hidden Universe

As we delve deeper into the mysteries of the universe, our understanding of the hidden realms of dark matter and dark energy continues to evolve. The journey to unravel these enigmatic forces takes us on an incredible adventure through the vast expanse of space and time. In this subchapter, we explore the exciting possibilities that lie ahead in the continued exploration of the hidden universe.

Particle physics has played a crucial role in our quest to comprehend the invisible components of the cosmos. Scientists have made significant strides in detecting and studying dark matter particles, although their exact nature remains elusive. The existence of these particles, which do not interact with normal matter and can only be observed through their gravitational effects, holds the key to unlocking the secrets of the hidden universe. Advancements in particle accelerators and detectors provide hope that we will soon gain a deeper understanding of dark matter and its role in shaping the cosmos.

Furthermore, the enigmatic force known as dark energy poses an even greater challenge. Responsible for the accelerating expansion of the universe, dark energy remains a profound mystery. Efforts to detect and quantify this force have been met with limited success so far. However, with the development of more advanced observational techniques and instruments, we are poised to make significant breakthroughs in the near future.

The continued exploration of the hidden universe also promises to shed light on the fundamental nature of space and time. The quest to reconcile general relativity, our theory of gravity on cosmic scales, with quantum mechanics, our understanding of the microscopic world, remains one of the greatest challenges in modern physics. By exploring the hidden universe, we hope to uncover new insights that may bridge these two theories and provide a more complete understanding of the universe.

The future of particle physics and the exploration of the hidden universe hold great promise. Collaborative efforts between scientists, international research institutions, and funding agencies have paved the way for groundbreaking discoveries. As technology advances, we can anticipate the development of more sensitive detectors, more powerful telescopes, and innovative experimental techniques that will allow us to probe the hidden universe with unprecedented precision.

In this journey through the hidden universe, we invite everyone to join us in the pursuit of knowledge and understanding. The mysteries that lie beyond our current understanding beckon us to explore further and push the boundaries of scientific exploration. Together, we can unravel the secrets of dark matter, dark energy, and the hidden universe, ultimately leading to a deeper comprehension of our place in the cosmos.

Final Thoughts and Reflections.

Final Thoughts and Reflections

As we conclude this incredible journey through the hidden universe of dark matter and dark energy, it is time to reflect on the profound implications that these mysterious phenomena have on our understanding of the universe. Throughout this book, we have delved into the depths of particle physics and explored the enigmatic realm that lies beyond our visible universe.

One cannot help but marvel at the intricate dance between dark matter and dark energy. These elusive entities, which make up the vast majority of our universe, have remained hidden from our view for centuries. Yet, their influence is undeniable, shaping the very fabric of our cosmos.

For the audience of particle physics enthusiasts, this exploration has been a testament to the incredible advancements made in this field. The discoveries of subatomic particles and the theories that explain their behavior have revolutionized our understanding of the universe. Through the lens of particle physics, we have been able to glimpse into the hidden universe, unraveling its mysteries one by one.

But this book is not just for the niche of particle physics enthusiasts. It is for everyone, as the secrets of the hidden universe have far-reaching implications for us all. Understanding the nature of dark matter and dark energy can help us comprehend the origins of the universe, the fate of galaxies, and even the ultimate destiny of our own existence.

As we reflect on this journey, we cannot help but be humbled by the vastness and complexity of the universe. It is a reminder of our place in the grand scheme of things and the awe-inspiring beauty that lies beyond our comprehension.

The exploration of the hidden universe has also shed light on the limits of our current knowledge. There is still much to be discovered and understood. It is a call to the next generation of scientists, thinkers, and dreamers to continue pushing the boundaries of our understanding and uncover the secrets that still elude us.

In conclusion, "The Hidden Universe: A Journey through Dark Matter and Dark Energy for Everyone" has taken us on a captivating expedition through the hidden corners of our cosmos. It has opened our minds to the wonders of particle physics and the depths of the hidden universe. Let us carry these final thoughts and reflections with us as we embark on our own personal journeys of exploration and discovery, forever seeking to unravel the mysteries of our universe.

www.ingramcontent.com/pod-product-compliance
Lightning Source LLC
Chambersburg PA
CBHW052204150726
48002CB00003B/1112